Genius

Richard Trevithick's Steam Engines

First Edition 2013 © Philip Hosken

Published by
FootSteps Press
Altarnun, Cornwall, UK.

Printed by
Lightning Source UK Ltd
Milton Keynes MK11 3LW

Layout, design and typesetting by
Daniel Nanavati

Cover Design
Sean Croft

ISBN 978-1-908867-05-6

Genius

Richard Trevithick's Steam Engines

by

Philip Hosken

Trevithick Society, Cornwall, UK

Dedicated to the late Frank Masahiro Trevithick Okuno, a proud direct descendant of Richard Trevithick.

With Thanks:

Without the assistance of those who helped me with *'The Oblivion of Trevithick'* this book could not have been written.

To that list I would add my gratitude to Lance Dearsley for his patient attention to the manuscript.

Any errors you find are entirely mine, caused when I fiddled with the text after Lance had finished.

Author's Note

William Shakespeare's contribution to literature has been profound and his influence grows year by year. Nevertheless, there are those students of his work who are determined to credit it elsewhere; we are told there are already some 5,000 books on the subject. Whatever their reasons most of these writers implicitly assume that such eloquent, thoughtful writing would have been beyond what they see as a poorly educated, simple country lad.

While Richard Trevithick added nothing to the literary scene, his development of an independent power source for most of the Industrial and all of the Transport Revolution changed the world. His genius left us with the control of high-pressure fluids that is still used extensively in industry and in all nuclear- and fossil-fuelled electricity power stations. However there were those, including some fellow Cornishmen, who were unable to come to terms with his outstanding achievement, and insisted upon crediting it elsewhere. While they appreciated the outcome of his work they were loath to acknowledge that it was from a man of humble beginnings who lacked their education and social standing. Trevithick may not have had grace and refinement, but his brain and determination provided the world with the technology it had been seeking for over a thousand years.

The sad result is that historians and raconteurs have found it in their interests to ridicule Trevithick for his failures rather than to bless and thank him for his achievements. It is suggested that, should this book be used for educational purposes a copy of *The Oblivion of*

Trevithick is held for reference.

Its opening lines are:

'The Missing Years in the history of steam engineering endeavour are those between the conclusion of James Watt's extended patent on the 'steam' engine in 1800 and the dawn of the Railway Age in 1830.

During this vital time there was feverish activity in the production of steam engines as industrialists demanded the new, more efficient, compact, high-pressure engines, and manufacturers rushed to supply them.

In all this excitement few historians accurately recorded how the new technology had emerged. The history of steam engineering is left with a gap that has been filled by crediting Watt with something he did not do, and crediting subsequent engineers with having invented the engines they merely built.'

This is the story of Richard Trevithick, true inventor of the steam engine.

P.M.H.

Young Richard Trevithick.
Photo: The Science Museum

Chapters

Foreword   xi

Chapter One:
  Young Trevithick and his surroundings   1

Chapter Two:
  Previous engines and dreams   14

Chapter Three:
  Trevithick's revolutionary contribution to Mankind   17

Chapter Four:
  Trevithick, dreams to reality   22

Chapter Five:
  Jane Trevithick   30

Chapter Six:
  Trevithick's next locomotives   32

Chapter Seven:
  Treachery and Duplicity   41

Chapter Eight:
  Davies Giddy   48

Chapter Nine:
  Tunnel, train and the man from Peru   52

Chapter Ten:
  Trevithick's Legacy   65

Humphry Davy   68

Appendix   71

Index   74

Foreword

Anyone who believes that the Age of Motor Transport resulted from the invention of the internal combustion engine, and started in the early 1900s, is missing a whole century of transport history.

The Locomotion, or 'Red Flag', Act of Parliament that many people date about the late 1800s was enacted back in 1865, several decades before any motor car we can imagine. So what, we may ask, was running about in sufficient numbers on the roads of this country for Parliament to want to restrict its speed and activity? There were so many steam cars and buses that the railway companies feared for their substantial investments. The railways lobbied Parliament and sought support from parish councils who claimed that the steam locomotives were damaging their roads. This claim had little foundation, as the pounding hooves of horses were likely to do much more damage. Nevertheless, the combined power of the railway companies, in which many members of Parliament had invested heavily, won the day and the small, scattered steam vehicle manufacturers lost out. While France, the United States and other countries went on to produce a variety of lightweight, quick steam cars the manufacturers in this country, restricted by speed limits, developed the world's finest steam traction engines. Many people are surprised to find that the Red Flag Act only affected steam vehicles; it was repealed in 1897, just as the internal combustion engine emerged.

All those steam vehicles, in this country and abroad, large and small, and all the railway locomotives and steam ships that were developed alongside them, had engines based on the principles established by Richard Trevithick. So why, you may also ask, if the transport revolution depended on the genius of one person, is that person not better known? While acknowledging the limited travelling capability of Nicolas Cugnot's gun tractors of 1769-71, we must give the credit for the first successful self-propelled passenger-carrying road locomotive, and shortly after that the first railway locomotive, to Cornish engineer Richard Trevithick.

In the following pages we will discover why the name of someone so devoted to his dreams of locomotion and with a lifelong dedication to their fulfilment, should not be a household word on everyone's lips, as are the names of Brunel, Watt and Stephenson.

Richard Trevithick, unknown source

Chapter One

Young Trevithick and his surroundings

The inventor of the steam engine, Richard Trevithick will always find a warm place in the hearts of Cornish men and women wherever they may be: and by engineers and all others who seek after historic truth. Let us look at what must have seethed through the mind of young Trevithick. Stimulated by his desire to build a mechanical contrivance to replace the horse, and surrounded by discontented mine owners using James Watt's atmospheric engines and resenting his incessant demands for money, Trevithick's contribution to industrial history was faithfully documented in the pamphlets and history books of the time.

So why are we continually being told that James Watt was the 'Father of the Steam Engine' and George Stephenson invented the railway locomotive? The answer offers little credit to some powerful men of Trevithick's day and to subsequent historians who failed to seek out the truth. In the years that followed Trevithick's invention, while his skill and ingenuity were in demand all over the world, the story of his genius surprisingly began to fade. This is not a full account of Trevithick's life and his many achievements, that can be found elsewhere. But you will become one of the few who will learn the truth: how Trevithick and his family were deceived, and how Watt and Stephenson were given credit for his achievements which they never sought. It must be said at the outset that Watt and Stephenson were two exceedingly clever men, and they deserved credit for all they did. Neither was a party to the deception of Trevithick, and neither denigrated him in any way. That was done by fellow Cornishmen.

1

This true story of Trevithick challenges the myth and Victorian romanticism that has replaced reality. It is a pity that we lack any significant account of Trevithick's personal life. Francis Trevithick, his third son and his biographer, avoided the many opportunities he must have had to tell us something of his father's home life and family relationships. He must have had access to the sources used in this book but for some reason either did not seek them or decided not to use them.

Indeed, Francis went beyond the omission of personal facts that would have improved our understanding of his father's activities. Being proud of his father and one of the first to appreciate that for some reason he had not received the recognition that was due to him, he created a number of questionable anecdotes that might have been based on traces of truth. They are found initially in the two volumes of his book, 'The Life of Richard Trevithick with an Account of his Inventions', published in 1872, when Francis had retired to live overlooking Brunel's broad gauge railway lines in the Great Western Railway terminus at Penzance .

Other writers found exciting and often misleading material in the books of Samuel Smiles that is not found elsewhere. They used the stories to enliven their own pages. But they availed themselves of this free information without apparently giving any thought as to its veracity. The effect on Trevithick's place in history was disastrous.

Trevithick's Childhood Home

Penponds, Camborne, Trevithick's home in childhood.

Their writing enables us more easily to recall what Trevithick didn't do than what he actually achieved. For instance, they repeatedly say that he didn't stay at school or listen to his teacher; that he failed to work at his desk in a mine office; that he ignored the injunctions of James Watt's bailiffs; that he refused to stop working on high-pressure steam when Watt said he should have been hanged for what he was doing; that he would continually get into all sorts of scrapes, lose money and generally lose his way. He had no time for Henry Harvey's sensible ideas, or for those of Samuel Homfray or John Rastrick. While he was kind and convivial, he was not a good husband. He endeavoured to provide for his family, but when he had the funds he found alternative uses for them. He did not write to his wife for years at a time. And generally speaking, he had little interest in anything that was not concerned with his engines or other inventions.

So, what can we be sure of about Trevithick? He was born on the 13th April 1771 to Richard Trevithick senior and his wife, the former Anne Teague. His father was a respected mine 'captain', (a term used to describe the manager and chief engineer of a mine), and an agent for the wealthy Basset family of local mine owners, later to be known as the de Dunstanvilles. He was the only surviving son, so he was probably spoilt by his four elder sisters and adored by the two who followed him. With a mother and six sisters to cater for his needs in childhood he must surely have learned to take the attentions of women for granted, something his future wife was to discover.

None of those who have written about Trevithick's early life have failed to mention the comments reputed to have been made by his schoolmaster who described him as 'a disobedient, slow, obstinate, spoiled boy who was frequently absent and very inattentive'. While it is clear that he assessed Trevithick's faults as he saw them, the schoolmaster didn't label his pupil 'unintelligent'. Today we could perhaps describe Trevithick as having 'special needs', and our caring society would be likely

Trevithick's cottage today.

to cramp his genius. Limited as his education was, we must not think of him as a dunce. He could hold serious conversations and could read and write, abilities that must have eluded many of his fellow pupils. Although he missed much of his schooling by truanting, and had not been very attentive when he was there, he learned enough to be able to engage in copious correspondence throughout his life. He paid little attention to his spelling, something his son Francis would correct for him whenever he had the opportunity. And the young Trevithick's mathematical ability was outstanding. Francis tells how he challenged his schoolmaster to complete calculations as quickly as he could in his head. We will never know how he did this, but we do know that he was able in later life to solve problems that had defeated others.

Francis suggests that his father went to school in Camborne but his parents had moved to Penponds on the outskirts of the town and it is likely that he attended a school founded by Mrs. Percival just across the lane.

Trevithick's size, strength and ability gave him the status of folk hero among his fellows a very useful asset for someone who expected people to understand and do whatever he required of them. There are numerous tales of his physical ability which, although many are unbelievable as they exceed the ability of Olympic athletes today, must have had some foundation in truth. He was made captain of Stray Park Mine while in his late teens. Throughout his life his willingness to offer challenges reflected his total belief in himself and the engines he built. Apparently he did not contemplate failure, and with some justification; his complicated, futuristic inventions all worked, although one or two required the development of materials to ensure their efficiency.

Trevithick's father's devotion to his only son extended to commissioning a miniature portrait of him, something rarely done at the time. And when school ceased to hold any interest for the boy his father provided opportunities for him to work in the offices of nearby mines. This was a very short term solution and soon Richard was out wandering among the mine workings, fascinated by the engines and other machinery, carrying a tablet and chalk to note what he saw and to record his own thoughts. He was emerging as an exceptional inventor.

Trevithick was born into a time of radical change. As he grew he saw many engines designed by Thomas Newcomen, a Devon blacksmith and Baptist minister, being replaced by the more efficient Watt engines. Newcomen's engines had been welcomed in Cornwall from 1720 as the first significant means of pumping water out of the mines. Although this had enabled dozens of Cornish mines to remain operational as they became deeper, the cost of fuelling their engines was huge; in some cases it was crippling. The Cornish mine owners faced a difficult situation. While Cornwall had substantial deposits of non-ferrous metal ores deep underground, extracting them was demanding, risky, and sometimes unprofitable.

Newcomen and Watt engines, together with their boilers, had to be built as well as fuelled, but Cornwall had no coal and precious little iron. The

5

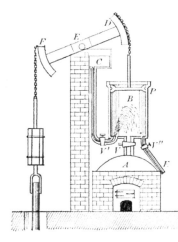

Thomas Newcomen's atmospheric engine, 1712 onwards.

Drawing: Wikipedia

mechanisation of the mining industry cost a lot of money. But profits were not available from the mines until the shafts had been sunk, the mighty engine houses built, and the engines and boilers had been designed, manufactured and installed. The boilers then had to be fed by coal brought from South Wales while the shafts were sunk to working levels. So a substantial quantity of capital had to be raised to start a mine. Prospective mine owners could not explore the ground by drilling, as is done today, and often relied on intuition and hope that riches lay in the depths below, a truly risky business. The people who did this were not surprisingly called 'Adventurers'. A few made a great deal of money but many failed disastrously.

A word must be said about the 'steam' engines that existed before Trevithick's time. The word steam has been applied to them retrospectively and incorrectly. The engines of Newcomen and Watt were known in their time as 'fire' engines, as their purpose was clearly to convert the energy from a fire into work but the steam generated by the fire did no work. It was condensed back to water to create a vacuum. The motive force was atmospheric pressure, which forced a piston into the vacuum. Atmospheric pressure is about 15 lbs/sq. inch, (1 bar at sea level) and was the maximum working pressure available to the Newcomen and Watt engine builders. Since working pressure could not be increased, the only ways of achieving greater power was the impractical increase of engine speed or the expensive increase in piston diameter.

A nerdy note must be made here about the frequent reports that Watt drove his low pressure

James Watt's condensing atmospheric engine, arrived in Cornwall 1777.

Drawing: Wikimedia

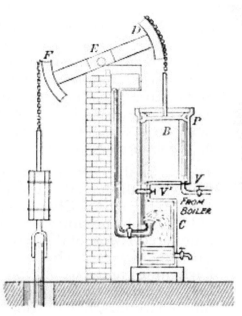

steam engines at about 3 lbs/sq. inch. 3 lbs/sq. inch is the notional pressure that people believe Watt generated in his boiler. Authors maybe correct in that assumption but the pressure was not used to drive the engine as they infer but to supply steam, as mentioned above, to create a vacuum. The operating pressure depended upon the atmosphere and the mechanical efficiency of the engine, it would have been about 15 lbs/sq. inch, still not a lot.

The Newcomen fire engine made deep mining possible, but it was grossly inefficient. A vertical cylinder containing a piston connected by a balanced beam to a water pump would, with some difficulty, be brought up to the temperature of boiling water and filled with steam. A small quantity of cold water was injected to condense the steam, leaving a vacuum below the piston. Atmospheric pressure above the piston then drove it downwards, often violently, causing the beam to rock and the rod attached to the other end to operate the water pump in the mine.

The weight of the pump rod in the mine then pulled the piston back to the top of the cylinder drawing steam in from the boiler. This splendid example of 18th Century engineering worked, but after each stroke the cylinder was left cold and had to reheated to boiling point before it would accept a fresh charge of steam. The need to produce sufficient quantities of steam to keep reheating the cylinder as well as to refill it to operate the engine was extremely inefficient. Many mine owners could barely afford to use these engines, and mines lurched between success and failure. The social consequences of mine failure were dreadful; a village would sometimes be made destitute overnight as the one mine closed on which all its inhabitants depended.

Coal and iron were delivered to Cornwall by a fleet of small ships that braved the dangerous westerly gales in the Bristol Channel. (Charts of areas such as St Ives Bay bear witness to the number of ships that foundered.) Most of these ships were operated by Cornish owners of what were often family businesses. Each mine engine would consume well over a thousand tons of coal a year and there were hundreds of mines. Small harbours, like the one at Portreath financed by the mine owners, sprang up along the north Cornish coast, but much of the coal and iron still had to be shovelled out of beached ships for distribution by pack horses.

In the late 18th Century the situation in the Cornish mining industry was precarious. Greedy entrepreneurs saw opportunities in winning valuable metal ores from deep below ground. Those involved in the manual work knew the dangers, but had to compete with each other for work in what was called the 'Tribute' system. Very often men risked their lives for less than a living wage.

When James Watt introduced his first pumping engine to Cornwall at Wheal Busy in 1777 Trevithick was just six years old. Watt's move to Cornwall brought him no joy. He wrote to his business partner Matthew Boulton telling him about an awfully wet place with broken engines inhabited by a "most ungracious" people. Clearly Watt, a sensitive Scottish engineer and inventor, was not cut out for business with strange,

James Watt, 1736 – 1819.
Picture: Wikipedia

unfriendly folk in a far-off land. It was a situation that would never substantially change as problems piled on each other. Among Watt's first acquaintances were the Hornblowers, a family of devout Baptist mining engineers at Chacewater, who had been building Newcomen engines throughout Cornwall (and, incidentally, the first one to reach North America). Their skills were useful to Watt and he even joined them at daily prayer in their little chapel. A friendly working relationship was established, and continued until Watt discovered that one of the Hornblowers had designed, patented and built a two-cylinder fire engine in defiance of his own patent. Watt turned to litigation while the Hornblowers continued to build and sell their own engines.

As news of Watt's new improved engines swept Cornwall he received a number of orders including one with some misgivings from Trevithick's father on behalf of the Bassets. The purchase of a Watt engine was a complicated affair. Each engine was specifically designed for the requirements of the mine. The owner would purchase the design and a licence to build the engine. He would also have to build a substantial granite engine house. A few of the components came from Boulton's factory; many were made by blacksmiths in Cornwall. While this presented opportunities for forward-looking blacksmiths like John Harvey of Carnell Green, whose daughter Jane would one day marry Trevithick, the monetary conditions attached to the purchase of an engine from Boulton and Watt did not find favour in Cornwall.

Watt described his new engine as having "improvements to the fire engine of Mr Newcomen." He had seen the need to reheat the cylinder after every power stroke of a Newcomen engine as its Achilles heel and he redesigned it with a linked but separate (pistonless) cylinder surrounded by water where the steam could be condensed. A valve in the pipe between the power cylinder and the condenser cylinder would be closed after the power stroke, and more steam would be injected below the piston; this destroyed the vacuum, allowing the weight of the pump rod in the mine shaft to draw the piston back to the top of the cylinder. Re-opening the valve allowed the steam to escape to the condenser, thus re-creating a vacuum below the piston to enable another power stroke. Time and steam were not wasted reheating the power cylinder, so there was no pause between strokes. As well as being much more efficient, these Watt engines were larger, more complicated and consequently more expensive than the Newcomen engines but the prospect of higher profits drove the mine owners into making agreements with Boulton and Watt that they were soon to regret.

Not only had the mine owners built new engine houses and bought the more expensive engines; they also agreed to pay Boulton and Watt a premium equal to one third of the savings in coal that they were making by using the Watt engine with its condenser instead of the Newcomen engine. This brought about immediate disputes as tests were conducted to calculate the performance or 'duty' of each engine. Engineers took sides and disagreed,

leading to some disputes
ending in court battles.
Teams of engineers
on the side of the
mine owners included
Trevithick and Davies
Giddy from St Erth,
the latter being a local
academic and Member of Parliament. Watt, who was
physically frail and suffered from headaches and
depression, was never a man to argue these cases, nor
could he undertake heavy engineering work. After
a couple of years he brought in William Murdoch,
an accomplished Scottish engineer, to supervise the
construction and operation of his condensing engines.

Watt had been canny enough to protect his
design with a patent so worded that it granted him
a virtual monopoly on all engines that incorporated
steam in their operation; he even secured an extension
to his patent that would not expire until 1800. The
demands of Boulton and Watt, together with the
patent, were described by commentators at the time
as delaying the arrival of the true steam engine for a
generation. While Watt's design had advantages over
Newcomen's for Cornish mine owners it was not the
end of the road for Newcomen engines. Being much
cheaper to build and not involving any premium
payment, the Newcomen engine was attractive to coal
mine owners who had fuel in plenty to compensate for
its inefficiency. While some five hundred Watt engines
were licensed during the lifetime of Watt's patent,
about two thousand Newcomen engines continued to
be sold in this country and abroad.

This was a time of mixed fortunes for the
Cornish mine owners. The Watt engines had certainly
enabled mines to be drained effectively at greater
depths but market forces had also come into play. Their

mines were beginning to face competition from the developing copper mine on Parys Mountain, Anglesey. It was indicative of the tense situation that the Cornish owners blamed Watt's premium charges for making their copper uncompetitive against the Welsh product. In this culture of blame the geophysical features of the two localities were often overlooked. The Cornish copper ore had to be lifted from deep underground where water threatened the very lives of those working to win it. The Welsh copper ore was on a hillside where the rain water could simply be allowed to run off, and no expensive pumps were required.

People are seldom happy in business no matter how much money they are making, and the situation in Cornwall clearly provided all those involved in mining with something to complain about. Watt was not a happy man. His association with Boulton was not the blissful commercial arrangement that so many have imaginatively described. Boulton had inherited a factory and a deal of money from his father; he then gained a fortune and an estate when he married. Within a year of his young wife's death he controversially married her sister, and accrued more from her wealthy father's will; moreover, this young lady also came to him with her deceased brother's share! Boulton went on to borrow yet more money, and even mortgaged Watt's expected receipts from his mining premiums in Cornwall. He travelled a great deal around the grand houses of Europe selling his ornate ormolu ware and paid little attention to the operation of his extensive factory network. So close did he come to financial ruin that his other partners suggested he should close the business before it failed. Watt, whose partner before Boulton had been made bankrupt, was very stressed by this situation and his unhappiness undermined any hope for friendly relationships in Cornwall.

As a bystander with ideas about steam and

engines, Trevithick was well aware of the difficult times faced by the mine owners who used Watt's engines and paid him annual premiums. Watt and Murdoch knew of Trevithick as a mine captain but we are told they did not consider him to be a threat to their operations until he teamed up with another young man called Edward Bull. Like Hornblower, Bull had sought to build a somewhat different engine in defiance of Watt's patent but the partners obtained an injunction against him, stopping him building further engines. The injunction did not restrain Trevithick, who went to work at Ding Dong mine in west Cornwall and completed Bull's 'upside down' engine. There is probably some truth in the tale that Trevithick's workmen upended Watt's bailiffs over an open mine shaft when they arrived to serve another injunction. They did not appear again.

Bull and Trevithick celebrated the completion of the Ding Dong mine engine by travelling to Boulton's manufactory at Soho, Birmingham in 1799. A feature of Boulton's public relations was his open house where prospective customers and others could look around his works and be treated to lavish hospitality. It was almost certainly Trevithick's idea to make the foolhardy visit and it was typical of his self-confidence that he could see no fault in it. The arrival at Birmingham of a 6ft 2in mining engineer with a strong Cornish accent was likely to attract attention. It did, and Trevithick was served with an injunction by Boulton and Watt's lawyers. We are told by Andrew Vivian, who was working for Boulton at the time, that his cousin Trevithick was very annoyed and refused to eat his dinner until the smell of the hot food overcame his anger. It had been a good day for Boulton and Watt as they had also achieved a legal victory over the Hornblowers. They celebrated with a firework display while a disgusted Trevithick headed off with Bull to Coalbrookdale in Shropshire.

Chapter Two

Previous engines and dreams

We have examined the situation in Cornwall concerning mine pumping engines in the years immediately prior to Trevithick's contribution to steam power. The development of engines that could harness the mystical power of boiled water had attracted the attention of dozens of men during the past thousand years or so. It would be unfair to jump directly to Trevithick's engine, the latest significant step in engine development, without mentioning something of steam and the former aspiring inventors' successes and failures. Firstly, we must understand that most claims were little more than dreams. The lack of suitable materials and the necessary fundamental understanding of thermodynamics meant that very few of the engines were ever built and none operated satisfactorily.

All accounts of engine development involving steam start with an illustration of the aeolipile designed two thousand years ago by Hero of Alexandria. Whether Hero had the skill and materials to build the aeolipile we will never know. Scientists today still argue over it, but it does reveal an early awareness of the potential of steam pressure.

Aeolipile from the 1st Century AD

Denys Papin, a Frenchman who came to England and became a Fellow of the Royal Society, heated water in what we would call a pressure cooker and, after an unfortunate accident, invented the simple pressure-relief, or 'safety', valve. Many others including Isaac Newton and Edward Somerset, Marquis of Worcester, tried in vain

Denis Papin's steam digester, 1679.

to harness the power of steam. Somerset published a design that Thomas Savery, a Devon man, later patented as his own. Savery, among others, saw that considerable rewards would accrue to the inventor of a successful engine. But he clearly showed that his marketing skills exceeded his engineering ability when he offered the mining industry an engine he called 'The Miner's Friend' for lifting water by means of fire. Subsequent writers, almost without exception, herald his engine as the first significant one that used steam. None of them, or the Patent Office, seems to have noticed that the second stage in the engine's operating cycle, forcing cold water up a pipe by means of steam pressure, simply could not have worked; even if Savery could have safely raised the required pressure and not lost it through joints in the steam pipes. So we are left with Thomas Newcomen as the first to build an effective engine that used steam. Unfortunately for him, the extended patent on Savery's useless engine did not expire until 1733. Newcomen had to make arrangements with Savery over royalties; like Papin he is believed to have died penniless and is buried somewhere in London.

Fire and water feature in all these engines and it is the combination of the two elements that make steam. But steam is an invisible gas; it is water vapour you see coming out of steam engines and kettles. Steam only exists at temperatures in excess of 100 degrees Celsius and usually under pressure. A small quantity of water will expand to about sixteen hundred times its volume when it becomes steam, (at atmospheric pressure), and will continue to expand if heated further.

As steam is heated and tries to expand it exerts tremendous pressure on the insides of its containing vessel, sufficient to burst a cast iron container and send heavy lumps of iron through walls or over considerable distances. When the early containing vessels failed they usually did so with enough force to cause severe damage to the surroundings and injury, even death, to any bystanders. These forces were well known by the civil engineers (as they were then called) of the day but they did not know how to contain them safely or employ them. Unknown forces were frequently believed by the general public to be the work of the Devil, and steam, out of sight and under pressure in a boiler, was one of those forces. It is no coincidence that one of Andrew Vivian's relatives referred to Trevithick's 1801 Camborne locomotive as a 'steaming puffing devil'. Its replica is known today as 'The Puffing Devil'. Those who tried to control steam, often seen as doing the Devil's work, were widely shunned for dealing in the dark arts. Trevithick was one of these but it never deterred him.

Men knew that steam under pressure could be put to work but they were thwarted in their attempts to use it. Trevithick's contribution to industrial progress was to contain steam and control its use.

Chapter Three

Trevithick's revolutionary contribution to Mankind

Once the ideas surrounding steam containment and control started to develop in Trevithick's mind he could think of little else. In Trevithick we see a man with an incredible imagination who was completely absorbed by his steam engines. He was clearly very fond of his wife and family but his fascination was with the steam engine.

Today we have no difficulty in recognising Trevithick's steam engine. It is the ubiquitous cylindrical boiler that distinguishes it from other machinery, and that was Trevithick's major contribution. Following his introduction of the high-pressure boiler, it was possible to contain safely what he called 'strong steam'. With no fundamental change the cylindrical boiler has become the essential component in all pressure machines. Today we find it everywhere, in power stations and nuclear submarines. Its shape appears in airliner fuselages, road tankers and the tanks of gully emptiers.

The cylindrical boiler is so commonplace today that we almost ignore it, but

The Trevithick Society's replica of the 1801 Camborne steam carriage passes the statue of Richard Trevithick outside Camborne library in 2010.

David Collidge.

its original design must have required a great deal of careful thought. It is doubtful if Trevithick had studied the activities of his predecessors; he probably did not care. His thought processes must have occupied all his waking hours, and maybe his dreams, as he worked way beyond the understanding of other men.

The Trevithick Society's conceptual replica of the 1801 Camborne steam locomotive climbs Camborne Hill with its intrepid crew during the annual Trevithick Da celebrations.

Photo David Col

In order to develop his revolutionary engine he would have to create what we call today a 'proof of concept' model. As this was a first it would be far from easy. Because its design was quite unlike anything that had gone before its operation would almost defy description. Jane Harvey's brother-in-law was a skilled clockmaker called William West whose name is on the mechanism of the Hayle town clock over the Harvey offices. Trevithick called him in to build a model. In the beginning the new technology must all have been a complete mystery to West and we must be grateful that he managed to build some excellent working models, two of which still exist today.

West's skill and patience were exceeded only by Trevithick's enthusiasm. With no engineering drawings West was asked to construct something that had never been seen before: this was the cutting-edge technology of the day. Bit by bit Trevithick explained how the model was meant to work and how it should be built. As the two of them sat together he watched his dream of an engine to replace the horse approach reality.

William West model of Trevithick's idea.
Photo: Science Museum

His excitement must have been unbounded. A mining engineer with a considerable knowledge of engines, pumps and the use of steam, he saw no way in which the model taking shape in the clockmaker's hands would fail to work. Some people might have had misgivings about the operation of the engine but Trevithick had none; his was the vision of a genius.

The models built by West worked, and one was demonstrated in the de Dunstanvilles' kitchen Tehidy Manor. Lady de Dunstanville turned on the steam and the little engine ran across the kitchen tabletop. This was a significant day for high-pressure steam, locomotion, cheap power, Trevithick and mankind.

Significantly, Trevithick asked West to build engines that were self-propelled. It was insufficient for Trevithick to invent just another engine that could perform the tasks required by mine owners and industrialists. In his mind his engine would do things other people had not considered possible. His determination to add wheels was important and many of the stories and myths surrounding Trevithick would involve wheeled vehicles.

While many men had dreamt of using steam power, none had discovered how to. Some credit should be given to James Watt in this respect. He had given the matter some thought, but access to suitable materials with which to build the boiler had eluded him. After a disaster with a wooden boiler bound by iron straps he abandoned for life the attempt to harness high-pressure steam. He even said that Trevithick should have been hanged for what he was doing. When he retired he was so fearful of high-pressure steam he included a covenant in the deeds of his house requiring that it should never be approached by a steam carriage. He was determined that his engines should have no connection whatever with road or rail locomotives.

All Watt's engines operated at atmospheric pressure. They were cumbersome beam engines, immovable and quite unable to propel vehicles, either on the road or on rails. In no way was Watt's design connected to the steam engines we recognise today.

William Murdoch, who is often cited as having passed his ideas to Trevithick, built one or two working model steam locomotives that used high-pressure steam. While his engine was similar in principle to that of Trevithick, and one model was famously reported by Francis as being tried at night, his square boiler would never have contained steam under sufficient pressure to operate a full-sized locomotive. He had failed to create what would be the fundamental feature of Trevithick's engine: the cylindrical boiler.

For six months both inventors lived in Redruth so they are frequently cited by authors as neighbours or living within a stone's throw of each other. Even the powerful Trevithick couldn't throw a stone from Merton House to Cross Street. Anyway, Trevithick ha

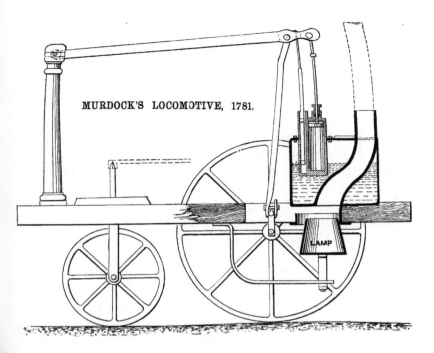

MURDOCK'S LOCOMOTIVE, 1781.

LAMP

William Murdoch's model locomotive, 1781. Richard Trevithick would have been 10 years old.

already done the development work on his engine prior to living in Redruth. Also, Murdoch's canny lack of communication about matters of industrial importance made it very unlikely that he shared anything with Trevithick. Francis Trevithick's story of Murdoch's steam engine trial in the church lane may have been imaginary to counter claims that his father had been advised by Murdoch.

But back to Trevithick and his cutting edge technology.

Chapter Four

Trevithick, dreams to reality

L et us trace Trevithick's initial achievement. Following, as he said, his inspiration one morning before breakfast, his wild dreams were taking shape in his mind. No one, except arguably Papin and Worcester who had used vessels similar to cannon barrels, had used any boiler which did not have at least one flat side. While some were square, the commonly used boiler was composed of riveted copper or wrought iron sheets in the shape of a 'haystack' or 'kettle' with a flat bottom below which a fire was lit. This was merely a vessel in which to boil water under no significant pressure. Sometimes such a boiler would fail to do even that safely, and many men were killed or injured including some who were working on the Boulton and Watt engines in Cornwall. Producing a cylindrical high-pressure boiler which enabled a locomotive to progress independently under its own power presented a number of hitherto unsolved problems.

Firstly, the lack of the usual flat bottom to the boiler would make it difficult to transfer heat from the fire to the water inside. Trevithick's solution was to place a furnace completely under water inside the boiler at one end. The furnace was made of an iron tube with a grilled floor. The heat from the fire was immediately transferred through the tube wall into the water. The hot fumes from the furnace travelled along a horizontal fire tube, still under water, until they reached the other end of the boiler. Trevithick's design then called for a U-bend to be made in the flue. This meant that the foundry men had to cut out several carefully shaped pieces of wrought iron and rivet them together to form a complete reverse bend. The tube continued back under the water to the furnace end of

An original riveted furnace and fire tube for a Trevithick engine made at Oak Farm Iron Works, Kingswinford. In Science Museum store at Wroughton.

Photo Science Museum

the boiler where it emerged to send the hot gases up a chimney. The furnace and tube had to resist the tremendous heat from the fire and remain watertight throughout. In addition, they had to withstand the high pressures being generated as the boiling water turned into steam.

With the earlier kettle boilers it was a easy matter to replace the water that had been turned to steam and consumed within the engine by simply opening a flap in the roof of the boiler and pouring in more. That is not possible with a boiler under pressure but, of course, replacement of the lost water is essential. To achieve it Trevithick designed a high-pressure pump that was operated from the motion of the engine; it forced water into the boiler through a non-return valve. The pump had to be capable of delivering more water than the boiler normally required; the excess was constantly recirculated to the reserve tank carried on the locomotive.

Previous engines had been built separate from their boilers. The transfer of steam from the stationary boilers to engines that were liable to rock when running tended to cause leakage from joints in the pipes. Trevithick's answer to this problem was to place the engine's cylinder and piston inside the boiler. While this presented a number of engineering difficulties there were several advantages that suited the operation of the engine and the eventual development of a locomotive. There was no loss of steam, pressure or heat. The unit was compact and the valves could easily be controlled from one place.

23

The overall insulation of this and other engines was not given the priority we might consider it should have been. Fire engines, as they were then called, were simply acknowledged as being hot and the effect of insulation on their overall efficiency would not be recognised for some years.

Four-way valve gear located on top of the cylinder allowed steam to act first on the top and then the bottom of the piston. Called 'double-acting', this produced reciprocating motion, which was not new but enabled the engine to drive a crankshaft. This separates its design from the single-acting engine that is more suited to pumping. The valve mechanism also provided an outlet for the steam once it had been used. Trevithick directed the hot, dangerous exhaust steam up the chimney.

Trevithick saw that the heat in the exhaust steam could be used to pre-heat the water used to top up the boiler. He therefore encased the hot exhaust steam pipe with a secondary pipe through which the boiler feed water was pumped.

Feeding the waste steam up the chimney had an effect that Trevithick may not have fully appreciated at first. Every time a blast of exhausted steam was released up the chimney it drew the waste gases sharply from the furnace and through the flue tube, causing fresh air to be drawn in through the base of

The modern welded furnace and fire tube made at Deepdale Engineering Co Ltd, Dudley, for the 2001 Camborne replica steam carriage

the furnace. This increased the supply of oxygen and significantly improved combustion in the furnace. Many men, including George Stephenson, claimed to have invented this blast pipe accessory to the steam engine over the next few decades, and one actually patented it. However, Trevithick clearly remarked on the advantages of this feature in a letter to Davies Giddy following the success of his plateway locomotive at Penydarren in 1804.

The usual course of engine invention is for a number of features to be designed, tried on existing engines and incorporated if they show any improvement. This usually involves a team of people in step by step progress and a number of disappointments. For instance, the principle of atmospheric pressure being used to achieve work was demonstrated by von Guericke in about 1650. It was developed by Papin, Savery and Newcomen. The industrial atmospheric engine was brought to its ultimate design by Watt and there it stayed. When its place in industry and mining was replaced by a completely new engine its technology had to stand alone and work first time or fail. All the necessary features appeared on Trevithick's engine and they all worked effectively and together first time. There was no sign of the usual step by step progress, this was an mighty leap.

Trevithick's work was revolutionary. None of the components in his engine had ever appeared on an engine designed by Watt or any other inventor. The leap forward in technology was immense. He had made a self-contained engine that was mobile, weighed a mere four tons, and could produce as much power as a Watt 'atmospheric' monster that weighed many times as much and required a massive granite building to house it. This was one-step miniaturisation that is difficult to find repeated even today. The cost of the engine had also been dramatically reduced. The

appeal of Trevithick's engine to industrialists was enormous.

From Trevithick's beginnings the *high-pressure* engine was exploited by numerous designers and engineering companies to power all manner of industry and transport.

By the time his patent expired in 1800 Watt had completed 23 years of his working life in Cornwall. He wasted no time in leaving the Duchy and dissolving his partnership with Boulton. He left the world of industrial power wide open for Trevithick to conquer. Stories of conflicts between these two men related solely to normal commercial practice. Murdoch had retired and left Cornwall a couple of years earlier.

The emergence of the full sized high-pressure steam engine owes a great deal to the interest shown in the contrivance by Trevithick's newly acquired father-in-law. In 1797 Trevithick's father had died, and later that year he married Jane Harvey. Her father, John, had already demonstrated his progressive thinking by establishing an iron foundry at Hayle.

About this time Davies Giddy enters the scene. Giddy's appeal to Trevithick was his understanding of the high-pressure steam engine, its potential and his willingness to help. He delighted the young inventor. Giddy was someone to whom Trevithick could turn, as he was able to answer his questions at length. This was probably a lot more than Trevithick wanted, a 'yes' or 'no' would have been sufficient. On the other hand, Giddy said elsewhere that he enjoyed answering Trevithick's questions; he was stimulated by the exercise of his mind.

One of the first answers Trevithick required was how much power might be lost from his engine if the used high-pressure steam was exhausted to air instead of being condensed as in the Watt engine. Giddy's answer was that the loss of power would not exceed one atmosphere (1 Bar or about 15 lbs/sq inch). Giddy reported that he had never seen a man as delighted as Trevithick when he received the news.

Giddy's contribution in the early years was not restricted to letter writing. He frequently travelled to wherever Trevithick was working and helped with experiments such as testing the feasibility of powering smooth wheels. He stayed with the de Dunstanvilles at Tehidy to await the conclusion of Trevithick's unsuccessful journey from Camborne during Christmas 1801 and offered considerable advice for Trevithick's first patent application. The complicated patent was completed in the names of Trevithick and his cousin Andrew Vivian early in 1802 and Giddy advised the pair to take it to London where they should seek the advice of Humphry Davy. They did so, but Davy was dismissive, passing them on to someone else. Long delays ensued during which Trevithick lost his patience and left Vivian to complete the application at his own expense. As a result, only Vivian's signature appears on the patent document.

The 1801 Camborne carriage had been completed as an outstanding piece of advanced technology. Sadly, it did not last long. An excited Trevithick drove it up Fore Street, Camborne (later to be known in the folksong as 'Camborne Hill') on a wet Christmas Eve. It lurched around the little town during the course of the next few days carrying Andrew Vivian and as many others as could hang on. Examination of reports on the events, including subsequent Government Select Committee Minutes, show that the steam carriage suffered from what we would call wheel spin. This raises the question whether the difficult line in the popular folksong called 'Camborne Hill' that goes, 'The Horses stood still and the wheels went around', should be, 'The Wagon stood still and the wheels went around'.

After all, there are no reports of horses being involved in the journey or in the neighbourhood. The well-known accounts of what happened on the way up the hill, supposedly written by Trevithick's men and published by Francis Trevithick, cannot be verified, and close examination raises doubts as to their validity.

The frequent suggestion that Trevithick climbed to the village of Beacon is also very doubtful. It is a long climb to Beacon and not something that would be considered possible by the team operating the replica of Trevithick's carriage today. Anyway, the road now known as Beacon Hill did not exist in Trevithick's day. There were few horse drawn vehicles in Camborne at that time, pack horses were used for the movement of freight. One can only imagine the state of the hill that Trevithick climbed.

A planned run to Tehidy to show the locomotive to Lord de Dunstanville ended in failure after a short distance. Later accounts of the event have been based on the one provided by Francis Trevithick; they have been embroidered over the years as romantic myth has been added to Francis's suggestion of what happened. While it is highly likely that the carriage went out of control and hit the wall at Rosewarne Manor, the story of an explosion has to be dismissed, and any reference to a fire of any proportion needs to be viewed with some scepticism. Irrespective of what Francis included in his book, we have the evidence from Giddy's journal for that weekend. He recounts his journey from Tehidy to Camborne, riding along the only road where Trevithick would have been travelling in the opposite direction. His destination was the Trevithick home in Fore Street, a little way up the road from the scene of the apocryphal fire, explosion and destruction. While he mentions that Jane Trevithick informs him of the Rev. Vivian's death, he does not mention having seen Trevithick, the carriage or signs of any problem.

Although the local papers record the happenings on Christmas Eve there is no account of anything untoward happening during the next fortnight. Eyewitness evidence given by Sir Goldsworthy Gurney of what he saw as an impressionable youngster watching with Andrew Vivian's son provides no report of any problems. Nonetheless, the various stories of Trevithick's failure are well ingrained in Camborne's folklore and will be repeated by storytellers and the media for many years to come. Whatever happened, the Camborne carriage was finished and Trevithick directed his attention elsewhere. It is a feature of human nature that we tend to enjoy the dubious rather than the factual.

We must remember that during this period of excitement Trevithick had numerous other responsibilities in the Cornish mining districts that also demanded his attention. We must not overlook another important character in this story: Trevithick's wife and companion, Jane.

Chapter Five
Jane Trevithick

Jane was the daughter of John Harvey, former blacksmith of the nearby village of Carnhell Green, who had seen a profitable future in the manufacture of iron pipes. We have John Harvey's letter to his wife, written as he sailed to Wales, and can compare it with a letter written by Jane to Giddy when her husband was in South America. While Harvey's letter is good for that of an 18th Century Cornish blacksmith, the beautifully executed letter by his daughter shows the manner in which he had educated his children.

Jane married Richard in 1797, the year of his father's death. The combination of two of Cornwall's respected industrial families was seen by all as a

blessed step, and happiness abounded. The couple set up home at Merton House in Redruth. They moved to Fore Street, Camborne after only six months and Jane soon discovered something about the man she had married; Trevithick failed to return the keys to the Redruth property and was

Jane Trevithick.
Photo property Barrie Osborn

Trevithick surrounded by her children in 1861. From left to right, Richard, Anne, John, Francis, Elizabeth and Frederick

harged a further six months' rent.

Outside his marriage, Jane's husband had one ll-consuming interest: his steam engines. He would pend most of their married life in love with them, urturing and protecting them to ensure they had a uccessful future. Meanwhile, Jane was left with the ouple's six children, trying to do the same for them, ringing them up much as if she were a single mother. he fact that she achieved this and managed in later fe to run the busy commercial White Hart Hotel in ayle reflected her dedication and hard work, and the elp she received from her brother Henry Harvey.

Jane's selfless commitment ensured that evithick could devote most of his thought and tention to his engines, something that would benefit e world. Her contribution should not be overlooked henever the story of Trevithick is told.

Chapter Six

Trevithick's next locomotives

Following the success with the 1801 Camborne locomotive, Trevithick's expertise was in considerable demand as mine owners and industrialists saw an economical engine that could be adapted to their needs. The possibility of boiler explosions was troublesome, but such matters were usually dismissed where there was money to be made; the factory owners installed the engines and made sure not to stand too close. Trevithick travelled throughout the country, explaining how his engines could be built. He developed strong relationships with foundries in Shropshire at Bridgnorth and Coalbrookdale. In 1802 the Coalbrookdale connection was to result in the construction of a steam locomotive that ran on tram tracks. Trevithick's enthusiasm for this engine is expressed in a letter to Giddy in which he explained how he was hoping to achieve hitherto unknown levels of steam pressure. Unfortunately, little is known about the fate of this important engine as the person behind the project died and his partners were not disposed to continue with the new fangled, possibly dangerous, contraption.

Things were different at Bridgnorth where a considerable number of Trevithick's industrial engines were built over the years together with the 1808 locomotive 'Catch-Me-Who-Can' that ran on a circular track in London. In the hands of John Rastrick the engine, known as a 'puffer' because of the noise it made, became much more adaptable. Study of one of his stationary engines in the London Science Museum will reveal that its parts can be assembled in several different ways. Such flexibility of design had not existed previously; it was a significant improvement

Trevithick industrial engine built by Hazeldine & Co, Bridgnorth.

over the engines built by Watt and others, which all involved expensive bespoke engineering. It reduced manufacturing costs and improved customer appeal. Trevithick clearly enjoyed his relationship with Rastrick and an examination of their correspondence reveals a remarkable understanding of each others' minds.

Trevithick was extremely busy on the Cornish mining scene while designing and building a steam carriage to operate in London. In 1803 his ungainly three-wheeled chassis left Cornwall by sea from Falmouth to have a carriage body built on it at Leather Lane, London. This vehicle was to become the first London bus as it plied for hire and charged passengers for journeys in the capital. Londoners did not exhibit the enthusiasm for his invention that Trevithick expected. His dreams were shattered, as were a row of railings when the carriage ran off the road. Clearly many years ahead of its time, the carriage was dismantled and its engine found employment at a barrel maker, leaving Trevithick to go off and deal with the many enquiries he had received for his industrial engines.

Plenty of mistakes were made as manufacturers produced the new high-pressure steam engines; but all sorts of machinery were being made and the pent up demand for power in industry was tremendous. It was clear that there was profit in making and using this new form of power. Soon skylines were dominated by smoking chimneys but the industrialists kept a safe distance from their boilers.

1803 was marred by what was possibly the only accident in which a Trevithick invention took lives. However machines are made, whatever safety measures are employed and no matter what instructions are delivered, there is always the human element. The boiler of a Trevithick 'puffer' engine employed as a pump at Greenwich, near the site of the current O_2 Millennium Dome, exploded causing the deaths of four men. It was subsequently found that the lad who had been left in charge had, with little or no appreciation of what the consequences might be, tied down the safety valve and gone fishing. Boulton and Watt, who had never used high-pressure steam and had always said that no one else should either, made a great deal of fuss about the explosion in an effort to prevent the sale

Tom Brogden's replica of Trevithick's 1803 London Steam Carriage

of Trevithick engines. Trevithick was very concerned by this and wrote to Giddy to say so; he also invented the fusible plug to avoid recurrences. When Watt's boilers occasionally blew up, (despite operating at insignificant pressures), it appears that little concern was shown for those killed and injured.

Generally speaking, Trevithick boilers were safe. There are no other reports of them exploding, whereas those made to other specifications often did. High-pressure steam engines were beset with sad stories and rumours. However, greedy industrialists saw profit in their operation, so sales continued. Trevithick demonstrated how safe a steam engine could be, in spite of his proximity to boilers all his life, he eventually died as an elderly man in his bed.

One of the enquiries Trevithick received was to develop into the most spectacular of all his achievements. Samuel Homfray, an iron founder at Penydarren near Merthyr Tydfil, where there was an abundance of iron and coal, had previous experience of atmospheric steam engines and, recognising the significance of Trevithick's invention, invited him to build engines in his foundry. It is easy to imagine the situation; after explaining to Homfray's workmen how to construct the new high-pressure steam engines Trevithick would have spoken of his vision of replacing all horses with his mechanical invention. Horses were used extensively around the foundry and for transporting the finished iron products to the Glamorganshire Canal nearly ten miles away at Abercynon. 'My engine can do a better job than those horses', he exclaimed. This typical Trevithick remark reached the ears of Richard Crawshay, a rival iron founder at nearby Cyfarthfa who did not believe the new-fangled contrivance could do any such thing. Homfray took Trevithick's side and bet Crawshay that Trevithick's engine would do anything its inventor said it could. The bet was a very substantial one: 500

Commemorative model of 1804 tram track locomotive on pedestal at Penydarren.

Photo Ray Jo

guineas or about half a million pounds in today's money; Trevithick had some very influential, wealthy friends but did nothing to take advantage of the opportunities they presented.

There was much activity during the following weeks. Trevithick kept in touch with Giddy by long letters advising him of the bet and how the construction of his tram engine was progressing. It has been assumed that the engine was a stationary one that had been adapted to run on tram tracks. As it reached completion and was successfully tested Trevithick asked Giddy to join him in South Wales; Homfray offered suitable accommodation. It was in these letters that Trevithick clearly described the improvement on the furnace of releasing the exhaust steam up the chimney of his locomotive as previously described. Giddy was in Oxford at the time and did no attend the first run to Abercynon on the 21st February 1804 but Trevithick wrote at length with a full account of what happened. He said, *'The publick untill now call'd me a schemeing fellow but now their tone is much alterd'*.

It was 25 years after Trevithick's triumph at Penydarren before George Stephenson produced his

Rocket and famously won the Rainhill Trials. Although dozens of locomotives had been built in the intervening years to operate freight services it was Stephenson who gained the right to produce locomotives for the first significant passenger railway.

Trevithick's account of the run to Abercynon revealed his excitement at the vindication of him and his locomotive. This was the world's first railway journey. It covered the significant distance of 9½ miles at five miles per hour, and the train consisted of five wagons loaded with ten tons of iron onto which a reported 70 people scrambled: an important event by anyone's estimation. However, there was a great deal of money at stake so arguments were bound to arise.

Homfray and Crawshay had both placed their 500 guinea wagers in the hands of another iron founder, Richard Hill of the nearby Plymouth iron works. The bet had been to convey the iron to Abercynon and successfully return with empty wagons to Penydarren, as could have been done by horses. Hill pointed out that the rules had not been properly observed. He noted that the tram tracks had been moved to the centre of a tunnel to enable the locomotive to get through, that the engine suffered a fault resulting in the loss of steam pressure on the return journey and had to be towed by horses. Also that the weight of the engine had broken many of the cast iron tracks.

While Trevithick acknowledged these points he only saw that his engine had done all that had been asked of it.

The locomotive went on to make similar journeys during the following days. Giddy arrived from Oxford on the 12th March, but we have to wait until he writes one of his final letters in 1839 before we have his account of the scene at Penydarren.

Even to this day trains do not run smoothly or

Trevithick's 1804 rail locomotive.

Painting by Terence Cun

regularly so we must give Trevithick credit for having made his locomotive work at all. The record of his achievement is overshadowed by stories of wagers, broken rails, track realignment and subsequent disputes. It is believed the wager was never paid and it's almost certain that Trevithick received nothing. However, we will see that he achieved fame from the events in South Wales, though it was soon to be taken away from him.

Homfray did not lose confidence in Trevithick' ability or in the future of his remarkable steam engine He offered his assistance and made an agreement to build and sell Trevithick engines, with two fifths of th profit going to Trevithick and three fifths to himself as the financier and risk taker. This worked and we read of Homfray being in dispute some years later with other manufacturers who were contravening

Trevithick's patent.

Today some sections of what should perhaps be the most hallowed length of railway track in the world are preserved. There are no traces of the tram rails but many of the original, or immediately subsequent, set stones that bore the rails are still in place near Merthyr Tydfil and a tunnel is open at one end. Much of it is pedestrianised and is known as the Trevithick Trail. A replica of Trevithick's contemporary 'Wylam', sometimes called his 'Gateshead', locomotive is in the Welsh National Waterfront Museum at Swansea and there are plaques to commemorate the event at both ends of the Trail. Few local people appreciate its true significance and not many railway enthusiasts make the pilgrimage to the birthplace of all railways. Abroad, sadly, it is barely known at all.

Study of Trevithick's actions in life does not reveal the shortage of temper so often suggested by writers. Short temper in a powerful man who stood a foot taller than his fellows would have been revealed by descriptions of bruises and destruction: there were none. But there is evidence that Trevithick would walk away from trouble or from people who refused to agree with his views on high-pressure steam engines. It was difficult in the early 19th Century to find anyone who understood how a steam engine worked. However, we have already seen that Trevithick was able to discuss the technology with his cousin Andrew Vivian, John and Henry Harvey, Davies Giddy, Samuel Homfray and John Rastrick. These people appreciated what he was doing and all were willing to offer him both commercial advice and, with the exception of Giddy, financial assistance. Many have expressed regret that Trevithick never had the assistance of a competent financier as several mistakenly see Watt's relationship with Boulton, but Trevithick had many such opportunities throughout his life. Sometimes he was unaware of their value, sometimes he was distracted by new developments. It is very unlikely that he would have found fellow feeling with Boulton whose character was too similar to his own.

The significance of the technological advances in steam engine development that occurred in Merthyr can be appreciated by considering that the people there witnessed the birth of the steam railway a year before Admiral Lord Nelson won the Battle of Trafalgar using ships built of wood and powered by wind.

Chapter Seven

Treachery and Duplicity

1804 had been a particularly significant year for Trevithick; he had built and driven the first railway locomotive in the world. However, before the year's end he was to receive a blow that would affect the rest of his life and the course of industrial history. Maybe it was a blessing he never found out.

A public demonstration of something exciting like the locomotive Trevithick operated at Penydarren is always assured of publicity and news of his steam engine, with its implications for transport and industry, swept the country. Trevithick was in demand everywhere. This is when he could have consolidated his achievements, developed the engine and built it to people's orders. Homfray offered his facilities and financial support for exactly that purpose. Some engines were sold by this means but Trevithick was soon off again, responding to enquiries that stimulated his mind and gave him opportunities to display the capability of his engine. As usual, money did not seem to worry him. Backers could see a substantial future in the development of his engines and his experiments were frequently funded if he simply asked for money. He was continually wishing he had more but when he had it he would spend it on the development of one of his ideas. We do not hear of him sending any substantial sums back to Jane who was struggling with his young family.

In October of that year Humphry Davy, who now was firmly established in London and had recently become a Fellow of the Royal Society, wrote to Davies Giddy advising him that Trevithick's invention was clearly going to be something significant and that he, Giddy, should take the credit for it. He said:

"Whenever speculation leads to practical discovery, it ought to be well remembered and generally known. One of the most common arguments against the Philosophical exercise of the understanding is 'Cui bono' [who will benefit]? It is an absurd and common place argument: but much used: so that every fact against it ought to be carefully registered. Trevithick's engine will not be forgotten, but it ought to be known and remembered that your reasoning and mathematical enquiries led to the discovery."

The powerful story of Trevithick's inventive genius continues but from here it is combined with an account of deception and misrepresentation that deprived Trevithick of recognition for his achievements.

The letter called for Giddy to accept credit for Trevithick's genius. The effect was immediate and damning for Trevithick. Davy's conspiratorial words were to stick in Giddy's mind and control his actions towards Trevithick for the rest of his life. What right did Davy have to determine the destiny and order the obscurity of one of the greatest inventors of all time? Why did Davy write that letter?

We will never know, but we can guess. The son of a destitute carpenter from near Penzance, Davy would always maintain that, in spite of the many claims of people like Giddy to have helped him, he had made his own way in life. And he had seen something the scheming Giddy had not. He worked in public while Giddy, a wordsmith, wrote in his study praising, encouraging or dismissing his many correspondents. Davy knew just how to achieve status in London society; he saw how the inventor of the steam engine was certain, despite some possible defects in his personality, to become important and famous. As word of his fellow Cornishman's success with the railway locomotive spread he would attract increasing media attention. He had every chance of becoming, like footballers and entertainers today, what we call a 'celebrity'.

Davy had met Trevithick in London when he arrived with his patent application for the high-pressure steam engine. He saw a man who was dedicated to invention; there was nothing he could do to restrain him. He probably didn't want to see Trevithick back in London, winning popularity that would have eclipsed his own. He may also have felt that fame would not sit comfortably on Trevithick's shoulders whilst Giddy, someone he knew, could trust and could probably manipulate, would embrace it. Giddy was now destined to wear the mantle of steam engine inventor, in his own eyes at least, uncomfortably and unconvincingly for the rest of his life. Why did he accept and act on Davy's argument? Why did he start undermining the man he claimed to have nurtured for years?

Giddy, an able politician as well as a writer, was immensely strong intellectually but always avoided direct confrontation. Davy was popular in London, his chemistry shows attracted large crowds and he was moving into the upper echelons of society. Giddy was well known in both administrative and regal circles; he had no wish to be disparaged by Davy. He recognised the danger of losing his hard won status at Court. Davy, an accomplished and insensitive showman, could have publicly trashed him.

It was a dilemma for Giddy. We can see from his subsequent actions that he decided to submit quietly and play down Trevithick's part in the history

43

of the steam engine. We know that his interest had previously been wholehearted. He had listened to Trevithick's ideas and offered advice. He had received his long letters and written at length in reply. He had attended his workshops, taken part in his experiments with traction and assisted in the preparation of his patent. Each man was a part of the other's life. Now all that had to change. But if he just walked away from Trevithick there would be no chance of influencing him and protecting his own position. He saw that he would have to remain Trevithick's confidant and stay close to him. If he was to make an effective claim to be the inventing genius himself, he had to know what Trevithick was doing. And Trevithick was still a young man; what other gains might come Giddy's way as a result of this deception?

Giddy said nothing to Trevithick about Davy's letter. He continued to receive Trevithick and answer all his correspondence. Trevithick never detected any change in the wily politician. On the other hand, never again did Giddy lift a finger to give Trevithick any practical help, or credit him with the invention of the sensational new machine. This suppression of Trevithick's achievement included the steam 'blast' pipe, a vital component that was to become the centre of many disputes among engineers who claimed to have invented it. When he received questions about the inventor of the steam engine Giddy would infer that Trevithick had merely built engines that he, Giddy, had devised. With luck, given the reputation he had already earned for himself as a man of science, he would get away with it.

In this way Giddy retained his position as confidant of the man he was deceiving, acquired news of his activities, and kept in with Davy and his friends in high places whilst advancing his own status in field where Trevithick was unlikely to tread. Giddy was to maintain this façade for the next thirty-five years, onl

Humphry Davy, source unknown

revealing the story privately in a letter to his son-in-law on his deathbed.

We have many instances of Giddy's change of attitude toward Trevithick. Within a fortnight of receiving the advice from Davy, Giddy was responding to a query about Trevithick from one John Trotter and devalued Trevithick in his reply by saying,

"I am not myself concerned in any Mine or business whatsoever and consequently I have never employed Mr. Trevithick or been connect with him in any undertaking. But having amused myself by investigations on mechanical subjects I cultivated his acquaintance as a sensible enterprising young man."

The following year we see Giddy in correspondence with William Nicholson about the steam blast pipe that Trevithick had made and then described to him in a letter from Penydarren. Giddy did not, as he could so easily have done, reveal Trevithick's involvement with the invention, probably hoping that Nicholson would mention him as the source of the information. However, Nicholson was as devious as Giddy; the following year he took out a patent for 'steamblasting': in his own name!

On one occasion Trevithick wrote directly to the Admiralty concerning one of his inventions. He sent the sealed envelope to Giddy and asked him to forward it to an appropriate person. Giddy did so, adding the following barbed note,

'My extraordinary countryman Trevithick has sent me the enclosed letter ... I presume it relates to steam and knowing Trevithick to be one of the most ingenious men in the world, although he has never done anything himself, I have complied with his rather singular request.'

At the time, Giddy was an important advisor to the government on mechanical and other matters. Using parliamentary language he had effectively invalidated the contents of Trevithick's letter.

Later, in a reply to John Wyatt in connection with the Thames tunnel project, Giddy dropped in the names of two people who had no relevance to the content of his letter,

"... I entertained for several years some acquaintance with Mr. Jonathan Hornblower; answering every theoretical question to the best of my power ... I had the good fortune to bring forward Mr. Davy earlier than without that assistance he could not have come to notice, and believing Mr. Trevithick to be a sensible, enterprising young man, I ten years ago began to lend him every assistance ... and his knowledge in mechanics ..."

Giddy then went on to recommend Trevithick as an engineer well equipped to drive the tunnel under the Thames, not because he was the inventor of the steam engine but because he was a 'mining engineer'. That was a challenge offering the prospect of a good reward and Trevithick took it.

In 1830 Giddy [or Gilbert as he had become in 1817] addressed the Royal Society at length on *'Progressive Improvements Made in the Efficiency of*

Steam Engines in Cornwall.' This was his opportunity to tell the assembled intelligentsia about Trevithick's achievements. After all, it had been twenty-six years since Davy had written his damning letter. Davy was long dead; mention now would have caused no harm. Gilbert simply could not face it; to declare that a lowly Cornish engineer had been the originator of the wonderful steam engine would have revealed his long-held dishonesty for having claimed to be its inventor himself. He would also have been asked why, as he had known Trevithick for several years, he had not brought this genius to public notice previously.

Gilbert was living a lie but so accomplished was he in the art of political manoeuvring that he managed to deliver the lengthy address to the Royal Society without a single reference to Trevithick. He lectured entirely upon the work of James Watt, who had retired thirty years earlier and been dead for eleven of them. Gilbert allied himself to Watt, whom he held to be the hero of the age, and, as he had done three years earlier in a lecture to the Royal Society entitled *'Observations on the Steam Engine'*, he skilfully avoided any reference to Trevithick or even Cornwall's lesser-known engineers, Hornblower and Woolf.

In 1831 Trevithick was asked to submit evidence to a parliamentary Select Committee on the use of steam engines to propel carriages but omitted some important references to his experience with the carriages he had made in 1801 and 1803. Gilbert was a member of the Committee and could have posed some leading questions but he did not.

Chapter Eight
Davies Giddy

Who was this Davies Giddy in whom Trevithick put his trust, who accepted that trust and then placed his own interests before Trevithick's but never revealed what he was doing? Giddy was a man of many parts. He was an academic who took an interest in a wide variety of subjects. He seldom spoke to more than one person at a time and, when he did so it was by means of a carefully written address. A profuse correspondent, he wrote to people at length, and each would receive dedicated attention to whatever subject they shared with him. However, he withheld information that might have benefited them if he thought it would benefit him more; in short, he was a control freak. Giddy was a powerful man who intended to remain powerful. He used all his wiles to hide a thirty-year association with Trevithick and only referred to him, when he had to, in the most casual manner.

Giddy was a man of many parts. Born the son of the curate at St Erth parish near Hayle in Cornwall, he lived in Tredrea Manor, a home inherited by his mother who has been described both frequently and mistakenly as wealthy. His journal contains an entry that the family sat during winter evenings round a small fire without a candle, and in 1770 his father was in receipt of alms from the parish. When his father was overlooked for the position of vicar he left the church to take up trading in tin and copper: he became very successful. Young Giddy was able to go to Oxford where he excelled and became known as the Cornish Philosopher. His interests ranged from the classics, literature, the Cornish language and the various philosophies to mining, mathematics and engineering in all its forms. He was also Member of

Davies Gilbert (formerly Giddy)
1767 – 1839

Parliament for Helston until the corn riots, represented Bodmin until the Reform Act of 1832, for a while was High Sheriff of Cornwall and also became a Fellow of the Royal Society, later to be elected its President. Such was his knowledge and willingness to participate in local Cornish activities that he was frequently consulted on mining and ngineering issues.

Giddy took an interest in gifted people and ied to advance their careers. One of these was homasine Dennis, a young lady from St Just, whom e instructed in the ancient languages, philosophies nd the arts. She has no part in this story except as an lustration of how Giddy would gain people's trust by king an interest in them whilst keeping his distance notionally. Giddy's intellectual prowess did not retch to understanding women. Thomasine pined for im and wrote a play in which he and she were clearly st as hero and heroine. After a disastrous episode her life when Giddy obtained for her the position of verness to the Wedgwood family, Thomasine nursed r tubercular sister and even slept with her, becoming fected with the deadly disease and so effectively took r own life. Giddy did not attend the funeral but later ected a tablet to her memory in St Leven church. The t that it was in Latin might say more about Giddy an about poor Thomasine.

Giddy took an interest in Jonathan Hornblower Chacewater, the engineer and steam engine inventor known to Watt. He advised him in connection with many technical matters, but Hornblower's stubbornness caused Giddy to lose interest and he turned his attention to the rising star: Richard Trevithick.

Although many people claimed to have discovered Humphry Davy, the young student doctor from Penzance who went on to become President of the Royal Society, Giddy's claim has some merit. He found a way of releasing Davy from his medical apprenticeship in Penzance and suggested him as an assistant to Dr Thomas Beddoes, a former reader in chemistry whom Giddy had met while he was at Oxford. Beddoes had established a remedial centre near Bristol with Josiah Wedgwood's money where he unsuccessfully attempted to treat sufferers from tuberculosis.

In time Giddy and his sister, Mary Phillipa, inherited Tredrea Manor and lived there together in a strong sibling relationship. Giddy was distraught when Phillipa decided to marry and, not being comfortable in the company of women, he lived a boring bachelor existence while at the same time evading the amorous approaches of Beddoes's wife, Anna.

Things changed when Giddy met and proposed to Mary-Ann Gilbert in 1807. The 32-year-old Mary-Ann was the impoverished daughter of Thomas Gilbert, a deceased grocer, living in Eastbourne with her mother. This is unlikely to have been a love match. Giddy had discovered that Mary-Ann was to benefit from the will of her childless uncle, a wealthy solicitor called Charles Gilbert who had considerable land holdings. To avail himself of this fortune he would have to marry the woman and wait until her uncle died. Giddy's determination to acquire a wealthy wife is revealed by his mortgaging of all the lands his father had accumulated in Cornwall to John Hawkins for £10,000 in order to provide a marriage settlement. Mary-Ann's uncle settled £12,000 on his niece, just enough in excess of Giddy's contribution to make a point. This was when a woman's property became that of her husband's on marriage but Charles had tied up matters to the benefit of his niece. Her uncle also wanted to maintain the family name and placed a condition in his will that whoever married his niece should adopt the name of Gilbert. Giddy married Mary-Ann when he was 44. Three years later Charles died, Giddy later became Gilbert and the couple inherited his fortune. Many people have murdered for much less.

The private life of Gilbert was far apart from his public life, where he was seen in Parliament and the Royal Society as a substantial participant in the country's affairs. Hitherto he had been in control of his entire life, but marriage changed all that. He seldom mentioned his wife in his copious correspondence. Once, he was asked about his marriage. His reply was that he thought, on the whole, it had been successful. Being comfortably situated Mary-Ann passionately wanted to help the less fortunate agricultural workers on her estate. She was generous to them and, indeed, to all about her. Gilbert encouraged his pupils in their studies but always drew the line at giving them financial assistance. He fiercely resisted the proposals then being made for the reform of the Poor Laws, believing that people should not be enabled by education to rise above their stations in life. The governmental control of human breeding, later known as eugenics, had a strong proponent in Gilbert. Life must have been difficult for the couple when fate delivered their first child, Mary, completely incapacitated; she died at the age of seventeen without ever 'giving sensations or motion from birth'.

Chapter Nine

Tunnel, train and the man from Peru

Trevithick's mind continued to burst with ideas. During the next few years he devoted much of his attention to marine matters. He mounted his engine on a boat for dredging the River Thames, invented a dry dock, demonstrated that water would not deteriorate if stored in iron tanks, and showed how precious space could be saved by using box-shaped iron containers in ships instead of the ubiquitous barrels.

Familiar as they are today, many of Trevithick's proposals failed to capture the imagination of the conservative Board of the Admiralty and others he approached at the time. Once, he was contracted to raise a sunken vessel off Margate by using the flotation of his iron buoys. He achieved this but refused to bring the ship to land as that was not part of the original contract. When his client refused to pay he cut the ropes and left, letting the vessel slip to the bottom again. This illustrates well Trevithick's singular method of dealing with disputes: peaceful and satisfying in the short term, but costly. A later idea involved using the recoil of a cannon so that it could be reloaded by only two men instead of the nine usually employed. Again the Admiralty were unimpressed. Through the years Trevithick was continually frustrated in his dealings with the Admiralty. They appeared to have had favourites among the engine builders, and he was not one of them.

Giddy's suggestion to the Thames Archway Company in 1805 that Trevithick was a suitable man to tunnel under the river at Rotherhithe fired the inventor's imagination with what he saw as simple work with the promise of a rich reward. Mining and pumping water were second nature to him and he saw no problem in driving a tunnel the few hundred yards from one bank to the other. Supported by a team of miners from Cornwall he commenced the task.

Digging through the soft clay a few feet below the bed of the river was a frightening experience for miners whose previous experience had been with hard rock. The tunnel Trevithick's men drove was low and narrow, sometimes opening out into chasms of quicksand. There was barely room for the men to pass each other and yet they had to transfer the spoil back through the whole length of the tunnel as they dug. Progress was painful but surprisingly quick. Trevithick was not popular with all the directors and the company employed surveyors to check on the progress of the tunnel. When it had exceeded 1,000 feet in length and with less than 200 feet to go one surveyor suggested that it was a foot or so out of line. This was of no consequence but it apparently riled Trevithick who drove a hole in the roof at low tide to indicate where he was. There was confusion as the tide rose and the miners tried desperately to staunch the inflow of water. They failed, and had to escape back through the tunnel to the other side of the river. Trevithick was last out and set off home to his wife and family at Limehouse.

Jane had been worried about her husband's activities in London and, against the advice of her brother Henry, had set out with their young family to join him. Away from the fresh air of Cornwall for the first time, she was appalled by the Great Stink, as the River Thames was then known, and soon left her husband's poor lodgings at Rotherhithe for something a little more acceptable at Limehouse. Trevithick was reported to have come in from his work hatless, without shoes, caked in mud and probably stinking, before tidying up and returning to the river.

Trevithick's Thames tunnel was never completed and, in 1808, he returned to steam

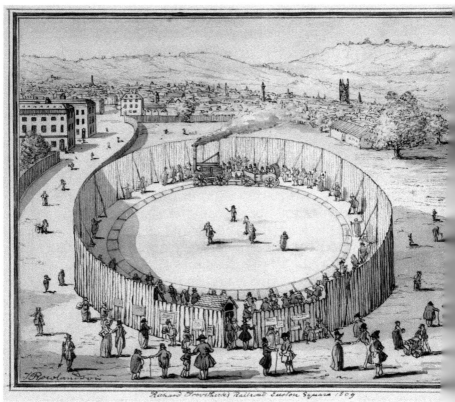

One of the several illustrations of Trevithick's 1808 'Catch-me-who-can' demonstratio were drawn long after the event.

Photo: Science

locomotion. His next venture was a circular rail track in the Euston area on which he demonstrated a puffer engine built by the Bridgnorth Foundry. Giddy said his sister had suggested the name of Catch-me-who-can for the locomotive. Trevithick charged for rides in an attached carriage and even offered to race a horse at Newmarket, but his hopes of finding a backer to develop a commercial railway were dashed.

The disappointed and exhausted Trevithick contracted a fever and sailed for Cornwall with his family, running the gauntlet of French naval ships. About this time his business affairs started to fail and he blamed his then partner, Robert Dickinson. Dickinson was a ne'er-do-well, so this may have been fair; it must be said, however, that Trevithick had willingly embarked on a number of ideas and ventures without any hope of remuneration.

Back home, Trevithick was made bankrupt. All his possessions in London, including a model steam engine, were seized by the bailiffs. But this depressing time of his life had its compensations; money from various ventures rolled in and Trevithick, too weak to spend it, was discharged from his bankruptcy by paying his creditors a commendable sixteen shillings (80p) in the pound. As he recovered in Hayle within the bosom of his family Trevithick received many visitors and plenty of well-intentioned advice. Had he applied himself to the development of his inventions and allowed his brother-in-law Henry Harvey to build them world engineering history would have been very different, but that wasn't his way of doing things. As he regained his strength, he visited the construction site of Plymouth breakwater with his friend John Rastrick of Bridgnorth and offered suggestions as to how a steam engine could be used there to advantage.

In 1811 a Swiss national called François Auguste Vuille but with the adopted name of Francisco Uville had been looking for a pumping engine to drain some of the silver mines of Peru. He had learned that the atmospheric engines of Boulton & Watt were unlikely to work satisfactorily at the considerable altitude of 15,000 feet (4,570m) and would be very difficult to transport to such a height. He then saw one of Trevithick's model engines, probably the one seized by the bailiffs, being offered for sale in London. He bought it and returned to the mines in Peru where he discovered that it worked perfectly well. Uville was a poor sailor but he returned to England in search of the engine's inventor, in the process rounding Cape Horn for at least the fourth time in his life. It is amazing what men will do when there are fortunes to be made.

Fortuitously, Uville landed in one of the

Fox packet ships at Falmouth, just a few miles from Trevithick's home. Soon the two men had met and discussed mine engines. Trevithick, excited to be back in harness, took Uville on a tour of Cornish mines, explained how his engines could be used to advantage in the Peruvian mountains and offered to have them made within six months.

A great deal has been written about Trevithick's life in Peru, revealing both his dogged determination to see though every project which attracted his attention and the manner in which he overlooked matters at home. The silver mines and wars of Peru, then pearl fishing in the Pacific and adventures in Costa Rica would occupy him for thirteen years. He started by redesigning his engines to make them suitable for shipment, transport overland and carriage up the mountains on the backs of pack animals. Until one is found, the actual design of these engines will be unknown.

We do know that six of the engines are reputed to have had 24 inch (60 cm) cylinders, much bigger than those incorporated in most of the familiar 'puffer' engines, and that these cylinders were designed by Trevithick to be assembled on site in Peru. Perhaps the engines were similar to the simple ones based on Trevithick's patent of 1802 and sold in quantity to sugar producers in the West Indies. The boiler plates were presumably riveted together in the mountains. We read that one boiler at the mine was successfully serving two engines, one for driving a pump and the other for lifting ore. How such a boiler could produce sufficient steam for both is something of a mystery. We do know that the boilers in Peru incorporated internal furnaces and flues for coal burning, but Trevithick made provision for external furnaces to supplement the internal ones as the only fuels available at first were wood and dung. Eventually poor quality coal was discovered in the locality.

The cost of Uville's requirements was more than he was permitted to spend, but nothing was going to stop Trevithick at this stage; speculating wildly, he offered to pay £3,000 towards the outstanding cost in return for one fifth of the mines' profit: a very speculative venture.

The engines were built by Rastrick at Bridgnorth with some components being manufactured elsewhere. The full shipment, including eight pumps with mining gear and a stamping engine for making silver coin, was loaded onto the 'Wildman' at Falmouth which, after some delays caused by the Spanish war in South America, set sail in September 1814. The shipment was accompanied by a small group of engineers including Trevithick's cousin, Henry Vivian.

The cargo must have occupied every inch of space in what was primarily a whaling ship. The perilous voyage around Cape Horn and northwards up the Pacific coast was followed by a cross-country journey of 160 miles (257 km) and a seemingly impossible trek up precipitous mountain tracks to the mines, where the Cornish engineers managed to put the machinery to work. The initial effect was remarkable; the mines were pumped dry and produced silver in abundance. But when the engines became unreliable for lack of expert maintenance Trevithick found the excuse he had wanted for going to South America himself, overhauling his engines and making his fortune in silver.

Trevithick then travelled to London to have the well known portrait of him painted by Linnell with a background of the Peruvian mountains. Whether he returned or not, he would leave behind a fine painting depicting him at what he probably saw was the zenith of his career.

Trevithick left Penzance in October 1816

aboard the 'Asp' with further 'sundry small' engines and a number of spare parts. On arrival in Peru he was heralded as Don Ricardo Trevithick and plans were even made to erect a silver statue in his honour. He soon had the machines working again and the mines producing large quantities of silver. But greed and politics led to disputes amongst the owners, and he left. Trevithick lost his personal stockpile of silver when the mines attracted the attention of warring factions from nearby Bolivia. His mining career in Peru was over.

However, Cornwall's interest in the enormous silver and other metal deposits in the Andes Mountains continued. In 1870 for example, Harveys shipped from Hayle an order for four engines with 37-inch cylinders made in 37 separate parts, no single one of which exceeded 300 lbs in weight.

While it is unlikely that Trevithick ever fully paid for the work undertaken for him by Harveys and other iron founders around the country, he left them with the knowledge and ability to make high-pressure steam engines. Among the many machines Harveys supplied for mining and other ventures all over the world were the largest steam engines ever made, those installed near what is now Schiphol Airport to drain the Dutch polders. (One of them has been preserved).

After leaving the Peruvian mines Trevithick is reported to have invested all he had in a pearl fishing boat, but little is known of the venture. For some years he sailed along the Pacific coast from the far south to Central America, intending, as he said after his return home, to make his wife the richest woman in Cornwall; but this dream, too, came to nothing, and he set off for home. During the latter part of his travels along the Pacific coast Trevithick had met and joined up with a Scotsman, James Gerard. The pair's struggle over mountains and through dense jungle did not occur in South America, as is frequently reported, but across

Haarlemmermeer pumping station, The Netherlands. Source unknown

Central America. On the way they were entrusted
with the well-being of the two young sons of a wealthy
Costa Rican coffee producer called Montealegre. The
lads shared the couple's adventures and eventually
reached Britain, one becoming a doctor in Scotland
whilst his brother returned to Costa Rica, eventually to
become his country's president.

Francis must have been privy to many stories
of Trevithick's adventures during his eleven years
of absence from Cornwall but tends to portray his
father as a swashbuckling hero. His description of
Trevithick's encounter with an alligator seems to
owe more to Defoe's *Robinson Crusoe* than it does to
reality. While this may have been due to Francis's
devotion to his father's memory, he could not hide the
fact that although Trevithick had written numerous
letters to financiers and engineers in England he had
failed to write to his wife. Jane, concerned for her
errant husband's wellbeing, and wanting to defend
him against allegations that he had acquired a second
family, wrote to Giddy for assistance. She wrote to
Trevithick, too, but whether her letter reached him we
shall never know.

On arrival at the Caribbean Sea the little

Fred Dibnah, MBE, explains Trevithick's achievement in Camborne

group of adventurers made their way south to Cartagena on the north coast of Colombia. Here something happened that is surely beyond the imagination of a soap opera writer; they met George Stephenson's son Robert, who was returning from an engineering trip to South America. Stories have it that he lent Trevithick £50 for his fare home, although the cost of the journey was nowhere near that amount. Stephenson took Gerard and the two lads to New York, surviving a shipwreck on the way and travelled on to Niagara Falls before eventually arriving back in England just a month after Trevithick. We do not know what Trevithick did during the Stephenson party's American tour, but he was unable to pay for his passage and skipped ship on arrival at Falmouth.

When he got home to Hayle he apparently walked in as though he had not been away. Francis tells how his father came to find him at school but gives no account of the greeting he received from his wife.

The arrangements Trevithick made to provide for his wife and six children during his absence seem to have been unsuccessful, as Jane relied heavily on assistance provided by her brother Henry. Henry set her up as manager of the White Hart Hotel he built in Hayle to accommodate customers visiting his foundry. Although unmarried, Henry and Grace Tonkin, his junior by 25 years, had nine surviving children. He and his then unmarried sister Betty, with whom he lived, also brought up the six orphaned children of another sister and her husband. Openly running two households and supporting a third, Henry was effectively father to 21 children, all of whom were well cared for and educated. Subsequent Harveys appearing in the story of the foundry and other operations were not his children but his nephews; his children were called Tonking.

Trevithick found the engineering world had progressed considerably since he left for Peru. His high-pressure steam engine was being copied everywhere and his family was receiving very little by way of royalties. Arthur Woolf was now the engineer at Harvey's foundry, where he turned the steam engine from a purely utilitarian machine into something of an art form. Woolf made much of his changes and added his name to them. While his only patent was seriously flawed theoretically, his claim to be an inventor served Gilbert well by enabling him attribute later developments of the steam engine to Woolf rather than Trevithick.

The Presidency of the Royal Society was held by Sir Humphry Davy from 1830 for seven years, and then by Davies Gilbert for three. The disturbing events during this decade are portrayed in *The Oblivion of Trevithick* and elsewhere. They illustrate well the character of some of the people Trevithick trusted. They were torturous years for the Society and were tactfully described by some of its Fellows as 'an unhappy period'.

Of course, Trevithick knew nothing of the internal machinations at the Royal Society and was unlikely to have understood them anyway. While he had been pursuing ideas for his engine his world in Cornwall had crumbled about him. Harvey was trying to resolve Trevithick's dispute with Woolf and his wife was clearly living an independent life; Trevithick saw little to keep him in Hayle. In February 1830 he set off for London and met up with his fellow adventurer James Gerard at Lauderdale House in Highgate where the two Montealegre lads had been lodging. He complained how the polluted air in London affected his chest.

Since his return to England Trevithick had resumed his correspondence and exchange of ideas with Gilbert as though nothing had happened while he had been away. Gilbert responded readily, resuming his unwavering duplicity. By this time Trevithick was talking about steam pressures exceeding 100 lbs/ sq inch (6-7 bar), and a new type of revolving steam engine that he referred to as 'werling'

John Hall, who operated a substantial engineering business at Dartford, (all too near London for Trevithick's health), saw opportunities in encouraging Trevithick and offered him facilities for building his turbine; Trevithick accepted the offer. He also developed his ideas on refrigeration at Dartford and in later years Hall & Co became a major supplier of industrial refrigeration equipment.

Trevithick's final years have been the subject of controversy in many accounts of his life, but as with his adventures in the Americas there are few records of his activities. Much of what has been published is the result of conjecture.

Although he was only in Dartford for a short time it is clear that Trevithick was well thought of there, as he is to this day. Hall paid him well and he lived in the Bull Hotel, the best in town, one that was patronised a few years later by Queen Victoria. It still exists and is now known as the Victoria and Bull. There are stories of him telling tales of his adventures in the Americas during the evenings.

Trevithick died at the Bull Hotel, of a chest complaint, on 22nd April 1833. The hotel keeper told his family he had been penniless and had owed him money, giving numerous authors cause to jump to the conclusion that he had died poverty

Richard Trevithick, statue erected in Camborne 1932.

stricken. In debt he probably was, but they are surely wrong about the poverty; he had some valuable items with him, was in good employment and was living in the best hotel in town. They have seen Trevithick's inability to capitalise financially on his inventiveness as a sign of failure. But very few financially successful people have ever invented anything and most inventors, certainly at Trevithick's level, seldom make much money. These writers have gone on to imply that Trevithick's employer was obliged to pay for his funeral and his colleagues had to form the cortege. They have been unable to believe that these acts might have been what the people of Dartford have always believed they were: final marks of respect by an appreciative employer and fond workmates.

We know that Trevithick's coffin was reinforced to make it proof against grave robbers and also that his grave was originally marked by a tombstone which disappeared, probably destroyed along with others. The location of Trevithick's lost grave attracted conjecture that continued in the local press from late Victorian times until well into the twentieth century. *The Oblivion of Trevithick* recounts all this but offers no further clues. It must be sufficient to say that Trevithick was not buried in a pauper's grave.

There would have been time for a member of Trevithick's family to travel from Cornwall for the funeral, but none was there. When Francis arrived at Hall & Company a few years later he was denied access to the factory. It was clear to him, however, that his father had been well cared for in Dartford and that his affairs, such as the making of his will, had been thoughtfully taken in hand.

Francis's biography of his father ends with an infamous passage, purported to have been written

Richard Trevithick, statue erected in Camborne 193?

by Trevithick himself, which begins with the words "I have been branded with folly and madness...." The passage has been continually repeated by other authors and the media without thought as to its provenance. They have used it because it offers what they see as an insight into Trevithick's life, a personal admission of his faults offering both confirmation of what they have already written and a neat conclusion to their books. Investigation shows that it was written by someone else, (probably by Gilbert as a final injury to Trevithick), that the original was much longer, and that Francis had altered it to favour his father.

Chapter Ten

Trevithick's Legacy

Trevithick was a single-minded man of huge energy and boundless inventive imagination. Freed in his prime from the stranglehold of Watt's all-encompassing patent, he devoted much of his life to the construction of the first viable true steam engines - and the first safe high-pressure boilers, on which they depended. His steam engines quickly exposed the limitations of Watt's atmospheric engines. All the engines and boilers which provided the steam power for nineteenth century industry and transport throughout the world were either Trevithick's or developments of Trevithick's designed by his successors. The availability of steam power led to production of consumer goods on a scale that even Trevithick could never have imagined. Sailing ships gave way to steam and eventually great liners opened the world to millions. Few people have done as much as Trevithick to advance mankind. He should be seen as one of the most influential figures in world history.

Of the man himself it could be said that he was big, strong, unconcerned about his own well being and single minded. He realised that high-pressure steam was the motive power of the future and he nurtured his engines as though they were his children.

Benefits of Trevithick's genius

Perhaps the most spectacular application of Trevithick's innovation was the Cornish mine engine. While his light-weight 'puffer' engines opened new opportunities by powering light industry and propelling transport by road, rail and sea, heavy industry and deep mines required ever more powerful engines. These were developed by linking Watt's beam engine and Trevithick's high-pressure steam boiler. By feeding beam engines with high-pressure steam whilst still using the separate condenser both efficiency and power output were greatly increased. Cornish foundries used the design to manufacture pumping and winding engines for the supply of domestic water and to pump water from mines. To this day Cornish engine houses, some complete with engines, are proudly preserved all over the globe and several engines still operate, on steam, at the Kew Bridge Steam Museum in London.

The railway locomotive became perhaps the most visible of Trevithick's inventions. His 'puffer' engine on wheels was developed to haul goods and passenger trains throughout the world. All the steam locomotives still at work on our 'heritage' railways are its direct descendants and, for a hundred years after Trevithick's demonstrations of road locomotion, steam was the sole form of power in heavy road transport.

The high-pressure steam engine was the supreme power force at the Great Exhibition of 1851 in the Crystal Palace, where it was seen as both a

Number 3 Machine Shop at Holman Bros, Camborne, manufacturers of high-pressure steam boilers and beam engines. The multiple belts operating the machinery were dr steam engines out of picture; must have been very noisy.

Trevithick Societ

workhorse and a work of art.
Mills and factories were using
steam power throughout the
world. The many steam engine
manufacturers included James
Watt & Co, successors to the late
Boulton & Watt. They adopted
Trevithick's high-pressure steam
designs, and only 25 years after his
death were installing boilers and
engines based on them in Brunel's
Leviathan, the s.s. Great Eastern. Trevithick had
changed the world for ever.

*£2.00 Trevithick coin struck by the
Royal Mint in 2004*

Photo: Royal Mint

And now ...

As this book is being written Richard
Trevithick and his locomotives appear on U.K.
television while the presenter heralds him, not Watt, as
the man who invented the steam engine that changed
the world.

Experiments are performed to show how
Trevithick's cylindrical boiler safely contained high-
pressure while others failed. It is unlikely that any
book will be written or programme produced in the
future about the rise of industrial or transport power
without due credit being paid to Trevithick's ingenuity,

What would Trevithick have made of this?
He probably would have taken it as the natural
progression of events he had imagined all along. He
may have been surprised, had he returned as this is
being written, to find his desire to replace the horse as
a means of transport had eventually come to pass. The
last traces of quaint, picturesque horse drawn wagons
have been outlawed in Romania. The poor creatures
are being consigned to the abattoirs to find their way,
labelled as beef, into meals throughout Europe he had
never heard of, such as beef burgers and lasagne. He
would have smiled with the rest of us.

Humphry Davy

Led by his brother John, a great deal has been written in adulation of Davy so there is little reason to repeat it here. Described as cantankerous he led a formidable life of lectures and research in London. He was in dispute with Michael Faraday who refused to undertake further research into electricity whilst Davy was alive for fear of his work being appropriated. He acquired numerous awards and titles including a knighthood, a few days after which he married a wealthy widow. For seven turbulent years he was President of the Royal Society and was followed in that position for three years by Davies Gilbert, a decade diplomatically described by a biographer of the Society as probably it most unhapp ten years. An experiment had gone wrong in 1812 and an explosion damaged his two

Statue of Sir Humphry Davy in Market Jew Street, Penzance.

Inset, the missing button

eyes, causing the loss of sight in one. Nevertheless, all portraits and his statue in Penzance depict him with two eyes. The sculptor quietly made his point by omitting a button from his waistcoat, this was intended to refer to his wife's lack of skill with a needle.

Appendix

Trevithick's inventive genius extended far beyond the steam engine and its applications mentioned in this book. Here is a brief summary of his patents.

1802
Construction of steam engines to drive steam carriages and other purposes (with Andrew Vivian)

1808
Machinery for towing, driving, or discharging ships or other vessels: steam tug (with Robert Dickinson)

1808
Stowing ships cargoes by means of packages: iron tanks (with Robert Dickinson)

1809
1.Floating docks: 2. Iron ships for Ocean Service: 3. Iron Masts: 4. Bending Timber 5. Diagonal Framing for Ships: 6. Iron Buoys: 7. Steam Engines for General Ships Use: 8. Rowing Trunk: 9. Steam Cooking (with Robert Dickinson)

1810
New Applications to propel ships to aid the recovery of shipwrecks; to promote the health and comfort of the mariners and other useful purposes (with Robert Dickinson)

1815
High Pressure Steam Engine; and application to useful purposes

1815
1. Plunger Pole Steam Engine: 2. Reaction Turbine: 3. High-pressure steam acting on water which acts as a piston: 4. the water from 3. used in 2. as in a Barker's Mill: 5. Screw Propeller

1816
A New Apparatus for evaporating water from solutions of vegetable substances

1827
New Methods for centring ordnance on pivots: Facilitating the charge of the same: and reducing manual labour in time of action

1828
New Methods of discharging ships cargoes and other purposes

1829
A New or Improved Steam Engine

1831
1. Boiler and Condenser: 2. Condenser in Air Vessel: 3. Surface Condenser: 4. Condensed water returned to Boiler: 5. Forced draught with hot air heated by condenser water

1831
A Portable Stove surrounded by water brought to boiling point

1832
Application of Steam Power to Navigation and Locomotion:
1. Super heater:
2. Cylinder kept in flue to be hotter than steam:
3. Jet Propulsion of Vessels:
4. Boiler and Super heater Applies to a Locomotive

Index

A

Abercynon 35, 36, 37
aeolipile 14

B

Basset 3
Beddoes, Dr Thomas 50
Bolivia 58
Boulton, Matthew 8, 11
Bridgnorth 32, 33, 54, 55, 57
Bridgnorth Foundry 54
Bull's 'upside down' engine 13

C

Camborne 4, 16, 17, 18, 24, 27, 28, 29, 30, 32, 60, 63, 64
Camborne Hill 27
Caribbean 60
Cartagena 60
Catch-me-who-can 54
Chacewater 9
Coalbrookdale 13, 32
Costa Rica 56, 59
Crawshay, Richard 35

D

Dartford 62, 63, 64
Davy, Humphry ix, 27, 41, 45, 50, 61, 68
de Dunstanville 19, 28
Dennis, Thomasine 49
Dickinson, Robert 54, 71
Ding Dong
  mine 13

F

Falmouth 33, 55, 57, 60
Fox 55

G

Gerard, James 58, 62
Giddy, Davies ix, 11, 25, 26, 27, 28, 30, 32, 34, 36, 39, 41, 42, 43, 44, 45, 46, 48, 49, 50, 51, 52, 54, 59
Gilbert 37, 45, 46, 47, 49, 50, 51, 61, 62, 64, 68
Greenwich 34
Gurney, Sir Goldsworthy 29

H

Hall & Co 62
Harvey, Henry 3, 31, 39, 55
Harvey, John
  Richard Trevithick's father-in-law 10, 30
Hawkins, John 50
Hayle 18, 26, 31, 48, 55, 58, 60, 62
Hill, Richard 37
Homfray, Samuel 3, 35, 39
Hornblower, Jonathan 46, 49

L

Limehouse 53

M

Margate 52
Merthyr Tydfil 35, 39
Merton House 20, 30
Murdoch, William 20, 21

N

Nelson, Admiral Lord 40
Newcomen fire engine 7
Niagara Falls 60

.

apin, Denys 15

Parys Mountain 11
Penydarren 25, 35, 36, 37, 41, 45
Penzance 2, 42, 50, 57, 68, 69
Peru ix, 52, 55, 56, 58, 61
Portreath 8
pumping engine 8, 55

R

Rastrick, John 3, 32, 39, 55
Redruth 20, 30
Richard Hill 37
Rosewarne Manor 28
Rotherhithe 52
Royal Society 15, 41, 46, 47, 49, 50, 51, 61, 68

S

Savery, Thomas 15
Schiphol 58
George Stephenson 1, 25, 36, 60
Stray Park Mine 5

T

Tehidy Manor 19
The Oblivion of Trevithick v, vi, 61, 63
Tonking 61
Tonkin, Grace 61
Trevithick, Francis 2, 21, 28
Trevithick, Jane ix, 28, 30, 31
Trevithick, Richard i, iii, iv, vi, vii, viii, xi, xii, 1, 2, 3, 17, 21, 50, 63,
    64, 67
two-cylinder fire engine 9

U

Uville, Francisco 55

V

Vivian, Andrew 13, 16, 27, 29, 39, 71
von Guericke 25
Vuille, Français Auguste 55

W

Watt, James vii, 1, 3, 7, 8, 9, 20, 47, 67
West Indies 56
West, William
  clockmaker 18
White Hart Hotel 31, 60
Woolf, Arthur 61
Wylam 39

Philip Hosken was born in Cornwall and educated at Truro School. He travelled a great deal in life to develop a number of companies and retired to Cornwall where he launched Cornish World magazine and edited it for six years. This was followed by his leadership of the project to build a replica of Richard Trevithick's 1801 road locomotive and an interest in the inventor's obscurity.

In 2011 he wrote *The Oblivion of Trevithick*, a weighty, original research into the lives of Trevithick and those around him. In that he revealed the deception that beleaguered the naïve genius and here, in *Genius, Richard Trevithick's Steam Engines* he publishes the salient facts of the case.

Philip lives in Cornwall where he is a bard of the Cornish Gorsedd who has travelled throughout the Cornish Diaspora. He is the chairman of the Trevithick Society and father to Treve, Tamsyn and Lowenna.

Lightning Source UK Ltd.
Milton Keynes UK
UKHW020620220519
343122UK00015B/881/P